Australasian Corrosion Association Annual Conference 2022

Corrosion and Prevention

Newcastle, Australia
5-8 June 2022

ISBN: 979-8-3313-0355-6

Printed from e-media with permission by:

Curran Associates, Inc.
57 Morehouse Lane
Red Hook, NY 12571

Some format issues inherent in the e-media version may also appear in this print version.

Printed with permission by Curran Associates, Inc. (2024)

For permission requests, please contact Australasian Corrosion Association Inc.
at the address below.

Australasian Corrosion Association Inc.
30a Jessie Street
Preston
VIC 3072
Australia

Phone: 61 3 8608 7900

https://membership.corrosion.com.au/

Additional copies of this publication are available from:

Curran Associates, Inc.
57 Morehouse Lane
Red Hook, NY 12571 USA
Phone: 845-758-0400
Fax: 845-758-2633
Email: curran@proceedings.com
Web: www.proceedings.com

TABLE OF CONTENTS

A FRAMEWORK TO PREDICT AND MONITOR THE DEGRADATION OF REINFORCED CONCRETE

A. Liso[1], E. Ariyachandra[2], A. Chegenizadeh[3], W. Green[2], M. Salasi[1], M. Iannuzzi[1]

[1]Curtin Corrosion Centre, [2]Vinsi Partners, Syndey, [3]School of Civil and Mechanical Engineering, Curtin University

SUMMARY: Reinforced concrete (RC) has been used with outstanding success in a wide variety of civil infrastructures. The in-service durability of RC structures can be undermined by the frequent exposure to aggressive environs triggering various deterioration processes. In particular, marine exposure conditions present a challenging environment for RC as a material. Despite decades of research, the underlying mechanisms of initiation and propagation of steel reinforcement corrosion and subsequent structural damage in marine environs remain elusive. This study aims to develop new laboratory protocols to monitor chloride-induced corrosion of RC and thereby to develop new automated corrosion monitoring techniques that are applicable for real-world structures. Multifunctional test specimens have been designed to assess key aspects of RC deterioration such as electrochemistry, chloride penetration (including diffusion and binding), and pore solution chemistry. The application of data generated by the protocols to develop a predictive model will also be explored.

Keywords: Reinforced concrete, Chloride-induced corrosion, Chloride binding, Pore solution chemistry, Laboratory test protocols

CASE STUDY: ANIMATION OF CHANGES IN PIPELINE POTENTIAL DUE TO TELLURICS

A. Verth, Verbrec, Brisbane

SUMMARY: Changes to telluric currents caused by disturbances in the ionosphere have long been known to impact the pipe to soil potential (PSP). Data logging of the PSP at pipeline test points allows the assessment of the level of cathodic protection to a standard. If we superimpose the data logger PSP at all test points over the same test period, we can observe the variation in PSP over the pipeline simultaneously. This can often reveal simultaneous anodic and cathodic sites as telluric currents enter and exit the pipeline. This static representation represents the speed at which telluric currents enter and exit the pipeline wall. By animating the pipeline potential over a short period (~2 minutes) and combining the PSP at all test points over the test period we can see the dynamic change in PSP. This observation can be used as a tool in identifying suitable recommendations to ensure the PSP meets the assessment criteria. Furthermore, in applying the use of this dynamic animation it is proposed sites along the pipeline length will be easier to identify for new cathodic protection installations. Similarly, other mitigation techniques including the identification of sites for grounding as a low resistance exit point for telluric currents can be identified)

Keywords: Potential, Pipeline, Telluric, Animation, Induced, Current.

AIRBORNE EM SURVEYS FOR LARGE SCALE PIPELINE CATHODIC PROTECTION DESIGN

A. Sterling[1], K. Hine[2]
[1]APA, Brisbane, [2]Mitre Geophysics, Brisbane

SUMMARY:

Pipeline cathodic protection relies upon quality ground resistivity measurements to allow accurate anode design. Traditional methods are time consuming with often inaccurate processes that simplify a complex earth into a simple 1D model. Ground access to do resistivities surveys is also difficult in many areas, such as cropped and fenced land parcels. Experience on the Roma to Brisbane Pipeline (RBP) is that some anodes, designed using this method, were ineffective.

Mitre Geophysics in conjunction with APA proposed and tested airborne electromagnetics for mapping ground resistivity along the RBP. The result represents a step change improvement in the pipeline operator's ability to plan and design effective anodes. The surveys are now being implemented Australia wide.

Keywords: cathodic protection, airborne EM, soil resistivity, pipelines.

THE MICROSTRUCTURAL EFFECTS OF LASER METAL DEPOSITION ON PITTING CORROSION OF AISI 316L, AISI 431 AND SAF 2507 STAINLESS STEELS

A. Hatem[1,2], C. Schulz[1,2], T. Schlaefer[3], N. Stanford[1,2], A. Roccisano[1,2], C. Hall[1,2]

[1]University of South Australia, Future Industries Institute, SA 5000, Australia, [2]Australian Research Council (ARC) Industrial Transformation Training Centre in Surface Engineering for Advanced Materials (SEAM), University of South Australia, SA 5000, Australia, [3]Laserbond Ltd, Cavan, SA 5094, Australia.

SUMMARY: This work compares the microstructure and corrosion resistance of additively manufactured AISI 316L, AISI 431, and SAF 2507 stainless steel samples fabricated by Laser Metal Deposition (LMD) technique. Equivalent wrought material counterparts were utilized as references. The corrosion behaviour was assessed by means of potentiodynamic polarization measurements and salt spray tests in a sodium chloride solution. The LMD materials presented differences in solidification mode and scale compared to its wrought counterpart. Samples presented small variations on the corrosion potentials and current densities. A comparatively wider passivation range was observed on all LMD samples. This was attributed to the marked effects LMD had on the microstructure. Salt spray test results demonstrated equivalent or superior pitting corrosion resistance of the additively manufactured stainless steel samples for exposure periods longer than 1000 hours.

Keywords: Laser Metal Deposition, Laser Cladding, Pitting, Electrochemical Corrosion, Salt Spray, Stainless Steel

QUANTITATIVE 3D CHARACTERIZATION OF CORROSION PIT PROPAGATION IN ADDITIVELY-MANUFACTURED SS316L SPECIMENS

A .E. Hughes[1,2], Jianli Li[3] ,Y.S. Yang[4], Majid Laleh[5], Haipeng Wang[3], Xufang Zhang[3], Jie Ma[6], Wei Xu[5], M.Y. Tan[5]

[1] Commonwealth Scientific and Industrial Research Organisation (CSIRO), Mineral Resources, Private Bag 10, Clayton South, Victoria 3169, Australia

[2] Institute for Frontier Materials, Deakin University, Waurn Ponds, Victoria 3216, Australia

[3] Institute of Theoretical Physics, School of Physics & Electronic Engineering and Collaborative Innovation Center of Extreme Optics, Shanxi University, Taiyuan, Shanxi 030006, China

[4] Commonwealth Scientific and Industrial Research Organisation (CSIRO), Manufacturing, Private Bag 10, Clayton South, Victoria 3169, Australia

[5] School of Engineering, Deakin University, Waurn Ponds, Victoria 3216, Australia

[6] State Key Laboratory of Quantum Optics and Quantum Optics Devices and Collaborative Innovation Center of Extreme Optics, Shanxi University, Taiyuan, Shanxi 030006, China

SUMMARY: *In-situ* micro *X-ray Computed Tomography* (X-ray CT) was used to investigate the effects of the growth of pits in additively manufactured stainless steel SS316L prepared using laser powder bed fusion (LPBF) 3D-printing process. A sample of 99.1% density was produced which contained lack-of-fusion (LOF) pore structures. To follow development of pit growth, the specimens were immersed in corrosive ferric chloride solutions and X-ray CT projection images were acquired periodically after every eight hours. Pit growth was quantitatively analyzed with a data-constrained modelling (DCM) method. It was found that pits initiated where the LOF pore sites intersected the external surface of the sample. Propagation proceeded along LOF structures, particularly at hatch boundaries and from there into the general matrix. Two different kinetics were observed: in the first case, pit volume did not change significantly with the time after initial formation. In the second case pits volumes increased with the time. The corrosion pit characteristic length (d=(volume)$^{1/3}$) versus time showed different power-law behaviors (t^{α}). At the lower end it increased as $t^{0.26}$ and at the upper end it increased as $t^{0.83}$. The lower and upper ends are likely to represent diffusion control (limited by the supply of reactants to the surface) and active corrosion conditions (salt film breakdown).

Keywords: Laser powder bed fusion (LPBF), Additive Manufacturing (AM), Selective-laser-melting (SLM), SS316L, corrosion, X-ray CT, data-constrained modelling (DCM).

CONSIDERATIONS AND CHALLENGES ASSOCIATED WITH MONITORING IMPRESSED CURRENT CATHODIC PROTECTION SYSTEMS

A. Cheaitani[1] & M. Cheytani[2]
Remedial Technology Pty Ltd, Sydney, Australia [1&2]

SUMMARY: Impressed current cathodic protection (ICCP) for reinforced concrete structures is a proven technology which has provided long-term corrosion protection to large numbers of bridges and wharves globally and across Australia.

One of the main factors related to the effectiveness of ICCP systems is the correct monitoring and maintenance of these systems. Ongoing adjustments of cathodic protection current over the life of a system based on the outlined protection criteria in global and Australian Standards is essential for achieving optimal long-term corrosion protection.

Some of the challenges associated with monitoring ICCP systems include accurate data retrieval for current adjustment from structures located in tidal areas and eliminating the risk of overprotection for prestressed concrete structures. For prestressed concrete structures subject to large tidal variations, the added complexity of monitoring necessitates the need for control systems with additional capabilities such as remote access and accurate data acquisition.

This paper presents operational data from ICCP systems installed on reinforced and prestressed concrete structures located in atmospheric and tidal zones. The data presented in this paper highlights key issues related to monitoring of such systems. These issues include the type of reference electrodes, the need for ongoing calibration procedures for embedded reference electrodes in concrete in certain applications, and the selection of control systems which allow for accurate data acquisition required for performing monitoring and adjustment of the cathodic protection system.

Keywords: Cathodic Protection, CP, Corrosion, Concrete, Prestressed, Reference Electrode

SYSTEM TECHNOLOGIES FOR CATHODIC PROTECTION OF REINFORCED CONCRETE

B. McGuiness[1], N. Critchley[1], O. Gasior[1]
[1]Marine and Civil Maintenance Pty. Ltd.

SUMMARY: A large variety of cathodic protection systems for the corrosion protection of reinforced concrete is available to asset owners and operators. This can make appropriate selection challenging for those responsible for concrete infrastructure. Our paper will review some of the most deployed systems with respect to performance, suitability and sustainability based on experiences with designing and installing both impressed-current and the newer part impressed-current, part galvanic, anode systems.

Keywords: Cathodic Protection, Galvanic, Durability, Design Life, Sustainability

PROPOSED SEEPAGE AND MOVEMENT MONITORING FOR AN 80-YEAR-OLD PLAIN CONCRETE RESERVOIR

B. Cunning[1], I. Chaves[1], L. Pilgrim[2], R. Petersen[1]
[1]Centre for Infrastructure Performance and Reliability – The University of Newcastle, [2]Civil, Surveying and Environmental Engineering Department – The University of Newcastle

SUMMARY: The investigation into serviceability and ultimate failure of ageing water retaining infrastructure is critical for cost, efficiency, safety and legality concerns. As a case study, this paper investigates an 80-year-old unreinforced concrete reservoir, where possible failure mechanisms are analysed to make an informed proposal for application of a suitable monitoring strategy. The proposed strategy considers appropriate equipment to monitor the serviceability and ultimate failure of the reservoir over a long period of time with careful consideration of the accuracy, cost, sampling rate and complexity of the network. Accordingly, strain gauge, tiltmeter, total station and piezometer measurements are proposed which require subsequent processing and finite element modelling.

Keywords: Reservoir Structural Health Monitoring

COMPARISON BETWEEN THE PERFORMANCE OF A CORROSION INHIBITOR BY DROPLET AND BULK VOLUME FOR AA6014

C Samarawickrama[1], Sebastian Pöhlker [2], Xiao-Bo Chen[1], Rou Jun Toh[1], P Keil [1,2], I Cole[1,3]

[1]School of Engineering, RMIT University, Melbourne, Australia, [2] BASF Coatings GmbH, Glasuritstrasse 1, 48165 Muenster, Germany, [3]Research Capability, RMIT University, Melbourne, Australia

SUMMARY: Structural aluminium alloys are widely used in aerospace, automotive and marine industry and this has led to it being extensively studied against corrosion. Corrosion protection of these alloys are most commonly achieved through protective coatings and corrosion inhibitors. With the ongoing necessity to discover comparable alternatives to toxic and carcinogenic chromate-based inhibitors, a substantial number of studies have been undertaken on determining the performance and mechanism of a wide range of organic and non-organic inhibiting compounds. The main focus of this study is to mitigate atmospheric corrosion through the use of corrosion inhibitors. Atmospheric corrosion is associated with the deposition of aerosols and fine droplets. Whilst majority of these studies have utilized well established methodologies for screening corrosion inhibitors in full immersion testing setups such as linear polarization resistance and weight loss measurements, these techniques are not suitable for evaluating corrosion under droplet volumes. In comparison to a bulk electrolyte environment, by utilizing a droplet volume, many factors such as electrolyte evaporation, secondary spreading, diffusion gradients, droplets volumes and shapes should also be taken into consideration. Therefore, this study focuses on a methodology developed to conduct electrochemical testing on electrolytes of droplet geometry to better represent atmospheric corrosion conditions.

Keywords: Atmospheric corrosion, droplets, corrosion, multielectrode system, electrochemical cell

LASER CLADDING OF FE-BASED ALLOY E-CLAD™ AS HARD CHROME REPLACEMENT FOR HYDRAULIC CYLINDERS IN MINING

C. Schulz[1], A. Roccisano[1,3], T. Schlaefer[2], A. Hatem[1,3], C. Hall[1]
[1]University of South Australia, Mawson Lakes, [2]LaserBond Ltd., Cavan,
[3]The Australian Research Council (ARC) Industrial Transformation Training Centre in Surface Engineering for Advanced Materials (SEAM), Mawson Lakes, Australia

SUMMARY: Laser cladding is a deposition welding process that can apply a defect-free layer of wear and corrosion resistant alloys onto a metallic substrate with final thicknesses ranging from 100 µm to several centimetres. In contrast to electroplated hard chrome, where the coating material is always chrome, laser cladding can utilize a wide range of metallic materials as protective coatings. This poses the possibility to adapt the coating to the service environment, for instance to design an alloy with tailored wear and corrosion resistance. While this presents an opportunity, it also makes the selection of the right material - a durable but economic solution - somewhat harder. In this study, a cost-effective Fe-based hard alloy, E-Clad™, was tested against the CoCrMo-alloy Stellite® 21 and the martensitic stainless steel AISI 431. The coatings resistance to sliding wear in a pin-on-disk setup and corrosion susceptibility via potentiodynamic scans in a sodium chloride solution were investigated. The proposed hard chrome replacement materials were benchmarked against an established hard chrome alternative, HVOF WCCoCr, and two different hard chrome coatings. The novel Fe-based materials were able to combine good wear and corrosion resistance while being an economic choice of coating material for hydraulic cylinders in mining.

Keywords: wear, abrasion, electrochemical corrosion, hard chrome replacement, mining.

ASSET MAINTENANCE – OPERATOR SITE SAFETY & TANK FARM PROJECT COST CHALLENGE

C. Reynolds[1]

BlastOne International

SUMMARY: Our presentation is about improving operator safety, removing risks of hidden costs & extending coating life in tank farms, using technology based inspiration.

Keywords: Operator Site Safety, Steel, Asset, Downtime, Protective Coating.

FAST RETURN TO SERVICE THICK FILM LONG LASTING LININGS – CHANGING THE PARADIGM

Dan Savage
DuluxGroup Protective Coatings

SUMMARY:

As is true with most industries, evolution, growth, and advancement in technology occurs to meet the demands of its customers. This is also applicable to coating technology and the application equipment used as well. While traditional thin film lining technologies have performed many needed functions asked of them over the years, I think we can all agree there are areas where these linings could perform better in one way or another to meet the demands of a modern world.

Today's think film coating technologies address many of today's demands, such as eliminating pot life concerns, being able to return to at tank to service quickly, help address safety and sustainability goals, and improved performance, leading to a longer service life.

Keywords: Application, Performance, Health, Environment, Rapid,

THE WEST GATE BRIDGE REFURBISHMENT

D. McKeown[1], [2]
[1]Dulux Protective Coatings, [2]

SUMMARY: The West Gate Bridge is one of Melbourne's and Victoria's most famous landmarks. Despite construction of alternate travel means to join the western suburbs and districts to the city, it remains a vital piece of roading infrastructure carrying between 80,000 and 100,000 vehicles per day. It has been open for over 40 years, and the original coating system on the steel box girder was due for refurbishment. This paper will chart the story of the refurbishment, but will also look back at the extensive planning carried out by the asset owner.

Keywords: Asset, refurbishment, steel bridge, maintenance, protective coatings

USE OF IOT IN REMOTE MONITORING OF CP IN MORE DIFFICULT LOCATIONS

D.H.Celine[1]
[1]Omniflex Pty Ltd, Sydney

SUMMARY: The "Internet of Things" (IoT) refers to the growing billions of things being connected together, all collecting and sharing data. Thanks to the invention of low power wireless technology and low cost computing, it is now possible to monitor more and more remotely located sensors and events. IoT technology is also finding its way into the world of corrosion monitoring and cathodic protection, providing the ability to monitor sensors and corrosion in locations where access to a power source and conventional means of communication are limited. This is providing valuable insights into rate of corrosion of structures and the performance of old and new cathodic protection systems that were not previously available.

This paper explores a number of use cases for this technology already in service and in development including the evaluation of the performance of galvanic cathodic protection systems over changes for example in tides and the weather, evaluating the efficacy of novel techniques of protection, the monitoring of individual anode currents in large cathodic protection systems fed from a single transformer rectifier, and the use of new satellite technology for the cost effective monitoring of extremely remote locations not serviced by conventional communications such as the mobile phone networks

Keywords: IoT, Remote Monitoring, corrosion monitoring, cathodic protection

MICRO-ENVIRONMENTS – PRACTICAL CONSIDERATIONS FOR RESIDENTIAL CONSTRUCTION

D. Harrison[1], P. Golding[1], A. Sheehan[1]
[1]Galvanizers Association of Australia

SUMMARY: The corrosion rate of zinc in Australia's macro-environments is well researched, with the various categories having been documented in the Australian Standard's AS/NZS 2312.2 and AS 4312. However, the effect of micro-environments created by poor construction practices can have a significant impact on durability and is not as well understood. In 2019, a study was started which aimed to investigate the effects of real time exposure at a severe marine site for a range of different steel sections, with various corrosion protection coating systems, using common construction practices for residential housing systems. A range of frequently used configurations were installed at the site over time, with each of the different corrosion protection options. This paper provides a progress report on their performance after 13-15 months exposure and discusses the impact of construction practices and design on localized corrosion.

Keywords: Atmospheric corrosion, micro-environments, galvanizing

METALLURGY AND CORROSION RESISTANCE OF 360 YEAR OLD CAST IRON PIPE FROM VERSAILLES PALACE

D. Nicholas[1], M.Dafter[2], I.Chaves[3], & S. Davies[4]
Nicholas Corrosion[1], Hunter H2O[2], Centre for Infrastructure Performance and Reliability, The University of Newcastle, Australia[3], Bureau Veritas Asset Integrity and Reliability Services Pty Ltd[4], Newcastle, Australia

Presenter email address: david@nicholascorrosion.com.au

SUMMARY: A sample piece from the original cast iron pipe system taken from the Palace of Versailles, France and installed in the 1660's has been examined and compared with a more recent cast iron water main material dating from the early 20th century. Metallurgical testing showed that the Versailles pipe sample, manufactured from a pre-industrial revolution charcoal blast furnace, was both harder and with a more complex morphology than a 'modern' cast iron pipe, and had significant quantities of minor alloying elements that were unknown at the time of manufacture.

Corrosion testing was carried out using Electrochemical Impedance Spectroscopy (EIS) in tap water to compare the general corrosion resistance of the Versailles pipe to the 20th century pipe. These results showed that essentially the corrosion rate of both materials was similar, and the minor alloying elements such as vanadium did not improve corrosion resistance. This is essentially the same conclusions as those drawn by Romanoff in 1957 when comparing early 20th century cast irons. It is concluded that soil type, installation techniques and backfill methodologies are more important factors in cast iron pipe longevity than material composition.

Keywords: Cast iron, corrosion, microstructure, Versailles Palace.

MORPHOLOGY OF A NEW TYPE OF COPPER CORROSION IN HIGH ALKALINITY POTABLE WATER

D.M. Nicholas [1],
[1]Nicholas Corrosion Pty Ltd

SUMMARY: Corrosion of copper drinking water pipes worldwide has been the subject of many investigations over the last 70 years and categories of corrosion type researched includes cuprosolvency, 'blue water', cold water pitting, hot water pitting, cavitation and corrosion/erosion. In the last few years, a new type of copper corrosion in cold potable water has been identified in high silica, high bicarbonate chlorinated ground waters in country West Australia. Corrosion is characterised by the formation of 'mounds' of copper carbonate on the pipe surface which stand proud of the copper substrate. Under conditions of high velocity and relatively high levels of free chlorine, an erosion mechanism causes 'thinning' between the mounds and subsequent high levels of copper release into the drinking water. This type of copper pipe corrosion has not been previously described in the literature.

The morphology of the phenomenon is shown, and possible mechanisms discussed, together with remedial treatment options for the water supply.

Keywords: Copper pipe, pitting, corrosion/erosion, copper carbonate

Recent and Current Initiatives of ACI Committee 222 on Corrosion of Metals in Concrete

D.G. Tepke[1], C. Halmen[2], and O.B. Isgor[3],

[1]SKA Consulting Engineers, Inc., Charleston, SC, USA; [2]University of Missouri, Kansas City, MO, USA;[3]Oregon State University, Corvallis, OR, USA

SUMMARY: ACI Committee 222, Corrosion of Metals in Concrete, has recently revised the Guide to Protection of Metals in Concrete Against Corrosion. Due to the significant impact to the industry, considerable variation in existing approaches and serious consequences of corrosion, the committee has also recently promoted initiatives associated with refining chloride limit recommendations for new construction and chloride threshold testing. This includes a recent Special Publication on chloride limits, revision of recommendations in the new guide, formation of a special Task Group, and sponsorship/oversight of CRC funded research for a collaborative study for the development of a standard critical chloride threshold test method. This paper will discuss some of the key updates to the Guide Document, as well as recent initiatives associated with chloride limits and thresholds. Information will be provided on the results of the CRC-funded research sponsored by the committee for a new test method for chloride limits. It will also discuss the reasons why these issues have been identified as important, impacts to the industry, and future needs.

Keywords: Reinforcing Steel, Concrete, Chlorides, Corrosion, Testing, Guide Document.

WHAT IS THE FUTURE OF ASSET INSPECTION?

G. Deutsch[1], D. Dibb[1], L. Holloway[2]
[1]Duratec Australia

SUMMARY: Currently asset inspections rely on experienced engineers to collect and interpret results from onsite inspections. This process can be time consuming and error prone but is crucial to knowing the structures condition, life expectancy and maintenance requirements. Comprehensive inspection manuals and technical organisations have helped ensure a high level of consistency and reliability across the industry in the past. However, these will never fully eliminate human error.

Unmanned aerial vehicles (UAVs) have provided a method of visual inspection which can be cheaper, safer, and quicker. Advances in high resolution photogrammetry has also made 3D photo realistic models possible, allowing users to remotely inspect a structures condition with a level of intimate detail that matches real life. Defect identification has been proven possible with artificial intelligence (AI) tools such as crack mapping on concrete structures or corrosion segmentation on steel structures. With advances in AI, UAVs, and modelling software, asset inspections will have fewer human errors, be performed more safely, become more repeatable, and become more efficient in delivery.

Keywords: Asset management, condition assessment, inspection techniques, UAV, artificial intelligence, structural health monitoring.

RELINING RAILWAY TANKERS FOR SULPHURIC ACID TRANSPORT IN AUSTRALIA

G.Small
Wulguru Rail Maintenance - Townsville

SUMMARY: This presentation provides an overview of the technical challenges that Wulguru Rail Maintenance (WRM) in Townsville had to overcome after award of a contract to reline railway tankers carrying sulfuric acid. Although the relining process is very common in the USA and had been done before in Australia, none of the parties with expertise were prepared to share it. Additionally, the coating manufacturer's Australian operation also had little to no experience with the specified lining system and could therefore also not provide substantial assistance. To make matters even more complicated, the owner and operator of the railway wagons has lost significant expertise with respect to the relining process around the time of award of the contract to WRM. And so started the journey to figure out exactly how to best install the single pack baked phenolic lining system on the internals of the railway tankers. From finding the right people to carry out surface preparation and coating application, designing and building the equipment to bake the lining between coats to meeting the client's quality management expectations, all within the tight budget constraints

Keywords: Rail transport, Sulphuric acid, Tank linings.

IN THE PIPELINE: AN ANALYSIS OF HYDROGEN EMBRITTLEMENT IN PIPELINE STEELS

G. Dour[1]
[1]Advisian (Worley)

SUMMARY: A literature review of the effect of gaseous hydrogen in steels was initiated in the early days of the hydrogen revolution. The intent was to understand the effect, if any, of gaseous di-hydrogen in steels. The review showed that dry di-hydrogen does have an impact, even if its dissociation into hydrogen protons is not a given. We found that the dissociation is active enough, in contact with the steel, to charge the steel with hydrogen protons. The diffusion of the protons is fast enough for them to reach the other end of a pipeline wall within hours. The potential concentrations are moderate, compared to electrolytic charging or sour gas charging. Severe degradation like Hydrogen blistering, Hydrogen Induced Cracking, Step Wise Cracking are not expected. Concentrations are however high enough to cause embrittlement, manifesting itself by a loss of ductility (or reduction of elongation to rupture), loss of fracture resistance (fracture toughness or energy) or higher fatigue crack growth rates. The review found that di-hydrogen partial pressure was an important factor as well as the microstructure of the steel. The paper insists on the latter as it should become an added requirement in the pipeline materials specifications for future use in the transport of hydrogen.

Keywords: Hydrogen, Hydrogen embrittlement, pipeline, Cr-Mn steel, API 5L, ferrite-pearlitic steel microstructure.

CHLORIDE DURABILITY OF REINFORCED CONCRETE MARINE STRUCTURES

H. Dyall[1], W. Green[1], R.E. Melchers[2]
[1]Vinsi Partners, Newcastle, Australia, [2]Centre for Infrastructure Performance and Reliability, The University of Newcastle Australia

SUMMARY: This paper presents the outcomes of an investigation of the durability of two reinforced concrete structures in the Port of Newcastle (58 and 113 years-old respectively) and sheds light on the future applications of Thermogravimetric Analysis (TGA) as a tool for estimating chloride binding effects for durability assessment of existing RC structures. Previous investigations of these structures have shown that many elements are in sound visual condition despite suspected chloride concentrations at the depth of reinforcement often being greater than 0.06% (by weight of concrete). In the present study, the same structures were tested by means of TGA of core samples and acid-soluble (total) chloride content analysis of dust samples. The laboratory results indicate that chloride binding has delayed the corrosion initiation process for these structures. TGA is shown to be a useful tool for qualitative comparison of the likelihood of chloride binding across different structures and structural elements but could not be used to quantify an estimated mass concentration of bound chlorides due to the inability to obtain an uncontaminated control sample from mature concrete.

Keywords: Chloride, Reinforcement, Marine, Thermogravimetric Analysis (TGA), Binding.

WELD CORROSION ON STEEL VESSELS AND OFFSHORE STRUCTURES

H. Humphrey, R. E. Melchers
Centre for Infrastructure Performance and Reliability,
The University of Newcastle, Australia, 2308

SUMMARY: Steel vessels such as ships and offshore structures are essential assets in the maritime commercial and defence industries. The corrosion of these structures exposed to marine environments has been identified as an industry issue for decades and remains so. This paper reviews relevant aspects of the existing literature focussing on general, pitting and weld corrosion, probabilistic modelling particularly for pit depths resulting from corrosion. The effectiveness of protective coatings and their expected life also are considered and it is noted that the available research literature is quite sparse, despite field experience reports for commercial ships and defence vessels indicating that weld corrosion and associated loss or impairment of protective coatings remains a significant issue for vessel life. An outline is given of current and on-going investigations for weld corrosion and coating life and the proposed application of both probabilistic modelling and finite element analysis.

Keywords: Steel Vessels, Weld corrosion, Probabilistic Modelling, Finite Element Analysis.

PRESTRESSED CONCRETE PILES CATHODIC PROTECTION PILOT

I. Godson, J. Heath, A. Czerwinski
Infracorr Consulting Pty Ltd

SUMMARY: A multi-span bridge has seven prestressed octagonal piles providing support to the reinforced concrete pile caps, where the piles are fully submerged in the marine environment. The submerged piles are suffering from alkali-silica reaction (ASR) instigated cracking, exposing the spiral ligature and prestressing strands to ongoing corrosion. As an alternative to long-term repetitive repairs by divers, a pilot cathodic protection (CP) system was designed and installed to one of the pile caps and the seven prestressed piles to control future corrosion.

The paper describes the pile arrangement and the basis of the Galvanic CP design and design assumptions, photographs and details of the installed pilot CP arrangement, including four aluminium anodes, negative connections and two reference electrodes. The CP system was connected to a remote monitoring system, with readings taken at two-hour intervals during the polarization stage, allowing the galvanic anode currents and instant off potentials to be recorded.

The prestressed reinforcement in the piles polarized to between -800 to -860 mV Ag/AgCl/0.5M KCl providing a safe level of protection and avoiding hydrogen embrittlement polarization levels of less than -900mV Ag/AgCl/0.5M KCl. The data obtained via the remote monitoring system permits the design assumptions to be directly compared to the actual system performance, allowing the design life of the system to be reviewed prior to the full-scale design for the remaining pile caps and piles of the bridge.

Keywords: Alkali silica reaction (ASR), prestressed, concrete, cracking, cathodic protection, hydrogen embrittlement

CONSERVATION OF A CORRODED 18 LB CANNON BALL FROM HMS SIRIUS (1790)

Ian D MacLeod
W.A. Shipwrecks Museum, Fremantle, Western Australia 6160

SUMMARY: The incomplete desalination of an 18-pound cannon ball from the wreck of the *Sirius* (1790) after two years of treatment combined with poor storage conditions on Norfolk Island caused significant post-conservation corrosion. Eight years after going on public exhibition the cannon ball cracked into many pieces. Originally the cannon ball had been part of the shot locker supply for the new wide-bore guns from Carron's ironworks in Falkirk, Scotland. Examination of the collapsed sphere of grey cast iron exposed a lozenge-shaped large lump of lead that had been inserted into the mould at the time of completing manufacture. Desalination had been effective for the first 3 cm of the 6.5 cm radius and had removed 84% of chloride ions. Detailed SEM examination of the residual microstructure showed the ball had typical grey cast-iron structure, with selective corrosion of the pearlitic phase to produce mainly magnetite which replaced the ferrite, while the liberated ferrous chloride mobilised into the surrounding concretion. The severely corroded inner core was rich in chloride, akaganéite and goethite. Enquiries indicate that the ball may have been an experimental piece of ordnance, cast in Scotland circa 1785. Theoretical treatment times for very large cannon balls and cannon have been developed to avoid future corrosion problems with other archaeological objects.

Keywords: Cast iron, Shipwrecks, Partial desalination, HMS Sirius, Cannon ball

LONGER-TERM CORROSION OF MILD STEEL IN NEAR-STAGNANT SEAWATER

I.A. Chaves, R.E. Melchers, R.B. Petersen, R. Jeffrey
[1]Centre for Infrastructure Performance and Reliability, The University of Newcastle, Australia

SUMMARY: Considerable research attention has been given in the past to both the short and longer-term corrosion performance of structural mild steels in open unprotected exposure conditions. However, the available information for the corrosion of mild steel exposed to seawater in enclosed, near-stagnant exposure conditions is limited. The present study reports data for corrosion loss of mild steel vertical strips exposed for 12 and 30 months within uPVC tubes and in near-stagnant tidal seawater conditions. These data are likely relevant for predicting corrosion losses inside windmill pylons used extensively in the offshore renewable energy industry. Estimation of the likely corrosion is of interest since field inspection is problematic owing to difficulties of access. The results show that the although the strips are subject to immersion tidal and splash zone corrosion there is no obvious accelerated low water corrosion effect and that after some 12 months exposure most corrosion occurs in the upper tidal zone and is significantly more severe when the local seawater has elevated concentration of dissolved inorganic nitrogen, indicating the involvement of microbiologically influenced corrosion. The likely practical implications of the present findings are discussed briefly.

Keywords: Steel, Seawater, Corrosion, MIC.

LONG-TERM MARINE IMMERSION CORROSION OF WELDED ABS GRADE HIGH STRENGTH STEELS

I.A. Chaves[1], Z. Sterjovski[2], R.E. Melchers[1], J. Rosen[1],
[1]Centre for Infrastructure Performance and Reliability, The University of Newcastle, Australia, [2]Defence Science and Technology Group, Australia

SUMMARY: Ship classification organisations typically publish requirements to standardise processes and product approval programs for materials manufacturers. For marine vessels including naval vessels high-strength steels are of particular interest because of their high cost to strength ratio. Corrosion performance of such steels has been considered previously, but data on the specific degradation of such steels when welded is scarce. This paper reports research outcomes for 2.5 years of marine immersion corrosion pit depth data for several permutations of high strength steel grades and weld types. The results reinforce previous findings that deeper pits tend to occur at the weld joint adjacent heat affected zones, irrespective of whether submerged arc or gas metal arc welding is used. The data obtained will be used to model corrosion loss and pit depth as functions of exposure period and hence can be used to possibly pre-empt joint failure due to high cross-sectional losses. The implications for the manufacturing industry are discussed.

Keywords: Steel, Seawater, Welds, Corrosion, Ships.

CORROSION UNDER STRESS: THE DEVELOPMENT & APPLICATION OF CORROSION CONTROL SOLUTIONS FOR PRESTRESSED CONCRETE STRUCTURES.

J. McLean[1]. J Heath[2]. A Rangott[3]

[1]Freyssinet Australia, Sydney, [2]Infracorr Consulting, Melbourne, [3]NSW Ports, Sydney

SUMMARY: Corrosion of Prestressed concrete elements can present substantial challenges for asset owners and managers. Being particularly aggressive and insidious in certain conditions, as well as difficult to identify by traditional visual inspection; The identification, management and control of this type of corrosion is of utmost importance for corrosion practitioners to ensure the future serviceability of these structures.

In order to address these concerns, a holistic, whole-of-life strategy must be developed and implemented which will take into account the practicality, constructability and site conditions for the solution as well as any future maintenance requirements.

This paper focuses on the challenges faced for asset owners and the mitigation techniques available to corrosion control practitioners when confronted with this issue. This will be discussed through the lens of a turnkey hybrid anode corrosion protection project, recently completed at the Bulk Liquids Berth No 1 (BLB 1) in Port Botany, NSW. This paper developed from the perspective of asset owner, designer and contractor will discuss the varying elements of the project and how a proactive, turnkey solution was developed to provide a 50-year life extension to the asset.

Keywords: Cathodic Protection, Prestressing, Concrete Corrosion, Hybrid.

CONDITION ASSESSMENT METHODS ON BRISBANE'S MOST CRITICAL SEWAGE RISING MAIN

J. Nadem, A. Bardak, A. West
SMEC Australia Pty Ltd, Brisbane

SUMMARY: Pipeline failures in trunk water and sewer rising mains typically incur expensive repair and replacement costs and customer disruption. The number of failures typically increases as the asset approaches its service life and can cause significant environmental and property damage. SMEC was engaged to undertake a condition assessment of DN1370 mild steel cement lined "Eagle Farm Sewage Rising Main". This main is one of a series of critical Urban Utilities pipelines which convey approximately 60% of Brisbane's sewage towards Luggage Point Sewage Treatment Plant. The condition assessment involved undertaking a desktop assessment in order to develop a targeted test plan based on likely failure modes. The test plan included the use of indirect assessment technologies (coating defect surveys and soil resistivity), before the implementation of direct assessment technologies at points of interest (wall thickness testing utilising UT and BEM technologies). The results indicated that whilst there were many significant coating defects, no substantial corrosion had occurred on the steel pipe wall.

Keywords: Corrosion, Condition Assessment, Soil Resistivity, Wall Thickness, Coating Defect, Cathodic Protection.

High Temperature Hydrogen Attack (HTHA) in Oil Refinery Piping Weld HAZ

J. Turnbull[1], R.G. Metcalfe[1] and R. Costanzi[1]

[1]Bureau Veritas, Asset Integrity and Reliability Services

SUMMARY: This paper presents the results of an examination of a section of piping located between two (2) heat exchangers in an oil refinery. The piping transported hydro treated oil and recycled gas from a reactor. The mixed phase liquid contained lubricating oil, hydrogen and hydrogen sulphide as well as light hydrocarbons, ammonia and potentially chlorides. The piping operated at 27 bar and at 260°C to 330°C. The piping had been in service for 13 years, representing approximately 50% of its expected service life, when leaking was observed at a welded joint between an elbow and straight section of pipe.

Leaking observed at a weld in the piping between heat exchangers 1 and 2 after a service life of 13years was the result of through wall cracking located within the grain coarsened region of the heat affected zone, immediately adjacent to the fusion boundary (HAZ-GC). The through wall crack had formed as a result of the linking of small, intergranular cracks associated with veins of decarburisation. Formation of colonies of small, intergranular cracks was attributed to High Temperature Hydrogen Attack (HTHA) in which hydrogen that has diffused through the steel and accumulated at grain boundaries reacts with iron carbide to form methane (CH_4), which builds sufficient pressure to form a fissure. Intergranular micro-fissuring was also noted in the inter-critical region, but not in the recrystallized grain refined region of the HAZ.

The operating temperature range of 260°C to 330°C (500° – 626°F) and a hydrogen partial pressure of 23 to 27bar (334 – 392psi) placed the pipework in a region that traversed safe to unsafe region of the Nelson curve for Carbon steel in the non-welded or Post Weld Heat Treated (PWHT) condition indicating susceptibility to degradation by HTHA.

Keywords: Hydrogen attack, HTHA, Oil and gas, Residual Stress, Weld HAZ, PWHT

MICROBIOLOGICALLY INFLUENCED CORROSION OF A PLATE HEAT EXCHANGER

J. Turnbull[1], R.G. Metcalfe[1]

[1]Bureau Veritas Asset Integrity & Reliability Services

SUMMARY: This paper presents the results of an examination of cooling water return lines and cooling cells within a 304L/316L stainless steel plate type heat exchanger. The heat exchanger had been in service for approximately two (2) years when corrosion was observed within horizontal sections of the pipes as well as within the cooling water cells.

It was understood that the failures were randomly located on the heat exchanger plates, however were more severe at the outlet port of the plate and in areas where water flow rates were either low or stagnant, such as at locations where the plates touch or under gaskets. Examination of the internal surface of the cooling water cells and the pipes revealed the presence of discrete pitting with associated tubercles, and sub-surface tunnel type corrosion of the delta ferrite phase immediately adjacent to welds. The tubercles in particular were considered typical of microbiologically influenced corrosion (MIC).

The cooling water temperature within the heat exchanger was reported to be within the range 30-40°C and to contain chloride concentrations of approximately 400mg/L (as chlorine), while the supply water had been dosed with oxidising and non-oxidising biocide as well as scale inhibitors. However, it was also understood that untreated water used during hydro-testing/commissioning was likely not removed and remained in place until the unit entered service some time later.

Analysis of corrosion product within the pits and within the heat affected zone of welds on both pipe and plate samples indicated significant concentrations of sulphur and chlorine. The presence of sulphur suggested sulphur reducing bacteria as a dominant species, while chlorine is known to be corrosive to austenitic stainless steels, particularly at the concentrations and temperatures reported.

Keywords: Corrosion, Microbiologically influenced corrosion, MIC, Weld, Heat Exchanger, 316L Stainless Steel

COATING THICKNESS TESTS (DFT) – UNDERSTANDING COMPLIANCE

Justin Rigby
Remedy Asset Protection

SUMMARY: Dry film thickness (DFT) is an important aspect of coating durability and has an impact on the lifecycle cost of structures. DFT test methods are described in various Standards, together with a variety of acceptance criteria.

However, where coatings are applied manually or in the field the commonly measured variations in applied film thickness needs to be carefully balanced with the objectives (or perhaps more appropriately, the intent) of the specification rather than marking areas that don't comply with standard acceptance criteria for repair without any further consideration.

In this presentation I will describe:

- Differences in DFT test methods according to ISO, SSPC, ASTM and Australian Standards.
- Differences in acceptance criteria within these Standards and difficulties in finding compliance to nominated DFT values.
- Provide statistical representation of DFT readings on a real project and analysis of the potential for errant readings.
- A proposed more pragmatic approach to interpreting and evaluating extreme low and extreme high DFT readings.

The information in this presentation intends to benefit industry through efficiency in the cost of construction by reducing the need for non-conformance reporting and consequent repairs where this does not appreciably increase durability of the applied coating system..

Keywords: Steel, Corrosion, Protective Coating, Dry Film Thickness, DFT, Spray Paint.

RESIDENTIAL CORROSION – IMPROVING SUSTAINABILITY

Justin Rigby
Remedy Asset Protection

SUMMARY: The residential construction market is not aware of the corrosion protections utilized by our infrastructure markets evidenced by numerous construction failures.

As an industry we aim to limit the cost of corrosion to society however this segment of industry has been left behind. Likely due to uninformed Owners and Investors.

Corrosion in residential corrosion is often managed through the requirements of the National Construction Code (NCC) and other regulatory documents that align under the NCC.

This presentation will provide case studies of new construction from Tasmania and Melbourne where corrosion has cost Owners unnecessarily and will describe;

- How the corrosion protections offered in the NCC is deficient.
- How the development industry specify corrosion protection and explain why there needs to be a change.
- How the builders warranty and insurance clauses treat corrosion protection.
- How specifiers can easily improve corrosion protection in contract documentation.

This information aims to raise awareness of corrosion potential in residential construction and provide some history when seeking to improve construction regulations in Australia.

Keywords: Steel protection, Protective coating, Residential Corrosion, National Construction Code.

Modelling of The Growth of Marine Bio-Films of Metallic Alloys in Seawater in a Nano-meter Scale

K. Habib[1]

[1]Materials Science and Photo-Electronic Lab., RE Program, EBR Center KISR, P.O.Box 24885 SAFAT, 13109 Kuwait

SUMMARY: Experimental data of electrochemical impedance spectroscopy (EIS) were used for the first time to predict (model) the growth of the marine biofilm on the surface of two metallic alloys in seawater. The EIS data, i.e., double-layer capacitance (C_{dl}), of UNS 1020 carbon steel and stainless steel UNS S304 were used in this study. The C_{dl} values of the formed biofilm on the alloys were measured in seawater on a frequent basis. The total exposure of the alloys to the seawater was ranged between 90 days to 180 days. The carbon steel was found susceptible to microbiologically influenced (induced) corrosion (MIC). A mathematical model was derived to gradual predict the growth of the biofilm formation by correlating the obtained C_{dl} of the biofilm to the thickness of the biofilm formation. The mathematical correlation was derived based on taking into the account parameters such as; the dielectric constant of the formed biofilm, the dielectric constant of the seawater at two temperatures 18°C and 33 °C, the volume fraction of the seawater in the biofilm, the dielectric constant of corrosion products, i.e., ferrous oxides; FeO, Fe_2O_3, magnetite; Fe_3O_4, iron oxide-hydroxide; FeOOH, Ferromanganese ;a mixture of MnO_2 and Fe_2O_3,Ferrous sulfate;Fe_2SO_4 , Ferrochromium; FeCr, in the biofilm, and the volume fraction of those corrosion products in the biofilm. The calculated thickness of the biofilms was in the range of a nanometer to a fraction of micrometer.

Keywords: Microbiologically influenced corrosion (MIC); Electrochemical impedance spectroscopy (EIS); Mathematical correlation; Seawater; Double-layer capacitance (C_{dl}).

ANALYTICAL METHODS TO ASSIST IN COATING FAILURE ANALYSIS

Kim Baker BAppSc(AppChem)

Paintspection

SUMMARY: The chemistry of protective coatings can often be overlooked or misunderstood when investigating a coating failure. A coating's chemical composition has been designed to provide it with optimal curing and physical properties. Errors or contamination during manufacturing and application can influence the coatings' ability to perform as specified. If a failure occurs, an investigator has the option to use a range of analytical test methods as part of their toolkit, to help provide scientifically proven causes for the coating defect. These methods can provide identification of coatings and their components, show contaminants in the paint or on the substrate as well as evaluate the integrity of the coating. Choosing the best methods of analysis can sometimes be overwhelming due to the highly technical nature of these tests. Understanding what tests are available, and the types of information they can provide, will assist investigators in finding the root cause of the coating's failure.

Keywords: Analytical analysis, Service Life, Inspector, Coating Failure, Mix Ratio, Remediation

SAND DEPOSIT THICKNESS AND ITS EFFECT ON UNDER-DEPOSIT MICROBIAL CORROSION

M. Diaz Mateus[1], E. Suarez[1], M. Salasi[1], J. Tarazona[1], A. Darwin[2] and L. L. Machuca[1]

[1]Curtin Corrosion Centre, WA School of Mines: Minerals, Energy and Chemical Engineering, Curtin University, Kent Street, Bentley, WA 6102, Australia.

[2]Woodside Energy Ltd., Perth WA 6000. Australia

SUMMARY: Physical and chemical characteristics of deposits found inside oil production pipelines play an important role in the mechanism of internal corrosion. Deposit thickness has been proposed as an important parameter that affects the ability of corrosion inhibitors or biocides to reach the metal surface and protect against corrosion. However, studies that correlate deposit thickness with microbial activity and the resulting corrosion damage have not been reported. In addition, so far, little research has been done on under deposit corrosion (UDC) with sand deposits accounting for the effects of bacteria in the field. In this study, we investigated the effect of sand deposit thickness on microbial activity and under deposit microbial corrosion (UDMC). For this, three different thicknesses of sand deposit from field (10 mm, 20 mm, and 50 mm) were evaluated on carbon steel immersed in field produced water containing microorganisms and corrosion inhibitor (CI). Different deposit thickness resulted in distinct microbial activity in deposits as well as dissimilar corrosion rates. Microstructure analysis revealed a morphology that indicated the development of a low pH condition under the deposits, which was in agreement with the metabolic profile of the field consortium. This research provided evidence of the corrosion risk on the internal pipeline surface exposed to different deposits thickness and bacteria. In addition, it provided insights into the microbiologically influenced corrosion (MIC) mechanism responsible for the damage. This study has direct implications for corrosion mitigation and highlight the importance of developing strategies concerning the removal of sand from pipelines where UDC and MIC are likely to have a conjoint effect on metals surfaces.

Keywords: Under deposit corrosion (UDC), Microbiologically Influenced Corrosion (MIC), sand deposit, thickness, carbon steel.

EFFECT OF SAND SIZE ON MARINE IMMERSION CORROSION OF MILD STEEL

M. M. Hossain[1], R. B. Petersen[2], R. E. Melchers[3], M. Bobby Kannan[4], I. A. Chaves[5]

Centre for Infrastructure Performance and Reliability, The University of Newcastle, Australia 2308

SUMMARY: Mild steel marine infrastructure is known to be prone to severe localised corrosion when in contact with coastal sands. According to established localised corrosion theory, points of contact of steel coupons and sand particles act as an anodic region due to partial or complete exclusion of oxygen. Contrary, surrounding portions of the sand particles act as cathodic region due to being more accessible to oxygen. Hence, the lower oxygen concentrated region undergoes corrosion. As sand size controls the ratio of cathodic area and anodic area, therefore sand size should have an apparent effect on corrosion. However, available evidences are anecdotal at best. Thus, research was conducted on marine immersed field exposures of mild steel coupons completely buried under different sizes of sands at the University of Newcastle Taylors Beach testing facility for up to a 24-month immersion period. Results indicated that the sand size hadn't much influence on mass loss unless some deep pits formation occurs for different reasons. Reasons and implications are discussed within.

Keywords: Mild steel, Sands, Field exposure, Corrosion, Mass loss, Deep pits.

THE EFFECT OF DIFFERENT MICROSTRUCTURES ON CORROSION INHIBITION OF HIGH STRENGTH STEELS

M. Behrouz Vaziri[1], A. Somers[1], B.R.W. Hinton[1] and M. Forsyth[1]

[1] *Institute for Frontier Materials, Deakin University, Burwood, Victoria 3125, Australia*

SUMMARY: The two most commonly used high strength steels in ship building industries are HY80 and DH36. These steels are considered as preferable alternatives to mild steels due to their excellent mechanical and welding properties. Despite the higher corrosion resistance of these steels in marine environments than that of mild steels, degradation due to corrosion still costs billions of dollars annually for marine industries. One of the most cost-efficient methods to minimise this corrosion is through the use of inhibitors. It has been reported that the combination of lanthanum and cerium with carboxylates can significantly enhance corrosion resistance of mild steel in chloride solutions. The performance of such inhibitors with other steel microstructures and compositions has not been assessed. In this study, the corrosion of HY80, DH36 and AS1020 mild steel in a control solution of 0.01M NaCl with lanthanum 4-hydroxy cinnamate was investigated. Electrochemical experiments together with optical microscopy, optical profilometry and scanning electron microscopy were used to study the corrosion mechanisms over 24 hours. For each steel, a reduction in corrosion with the addition of the inhibitor was observed, however the effect of the inhibitor for each steel was markedly different. HY80 with a tempered martensitic microstructure, showed the highest corrosion resistance in the control solution, but showed the highest corrosion current with the lanthanum 4-hydroxy cinnamate inhibitor present. The DH36 steel with a ferrite and pearlite microstructure, had a moderate corrosion rate in the control solution, but the effect of the inhibitor when present was greater than for HY80. AS1020 with a similar microstructure to DH36 but with less pearlite, had the highest corrosion rate in the control solution and the lowest corrosion rate with the inhibitor present. It is hypothesised that at the early stage of immersion the inhibitor molecules bond through the carboxylate group to active anodic sites adjacent to cathodic sites, such as the high carbon phases i.e., cementite/carbide and/or inclusions such as sulphides. With time this film builds up and the La hydroxide deposits at the cathodic sites to reduce their activity. This will form a protective inhibiting film, thus mitigating corrosion during longer immersion times.

Keywords: High strength steel, microstructure, composition, ferrite, pearlite, inhibitor.

CORROSION CONTROL TECHNOLOGY NEEDS FOR FUTURE FUEL INFRASTRUCTURE

M. YJ Tan, Y. Huo, F. Varela, I. Ubhayaratne and Y. Xiang
School of Engineering and Institute for Frontier Materials, Deakin University, Victoria, Australia

SUMMARY: The introduction of future fuels such as hydrogen into existing fuel infrastructural network brings up potential new threats such as hydrogen embrittlement, in addition to aging and corrosion, to assets such as pipelines and storage tanks. This paper provides an analysis of technological gaps, needs and opportunities for fuel infrastructure based on information acquired from a recent industry survey and literature review. Particular focus is on infrastructural integrity management and protection technologies such as smart monitoring and asset condition prediction. The detection, monitoring and prevention of hydrogen induced pipeline integrity issues such as hydrogen embrittlement, cracking and leaking have been identified as prime knowledge gaps for future fuel infrastructural networks, while "non-piggable" pipelines with disbonded shielding coatings have been identified as 'highest risk' assets in existing fuel infrastructural networks.

Keywords: Infrastructure, Pipeline, Integrity management, Hydrogen embrittlement, Corrosion monitoring, Protective coating.

COMPARATIVE EFFICACY OF SELECTED NON-OXIDISING BIOCIDES AGAINST OILFIELD MICROORGANISMS IN THE BIOFILM STATE

N. Leinecker[1], B. Tuck[1], S. J. Salgar-Chaparro[1], S. McKay[2], A. Darwin[2], J. Still[2], L. L. Machuca[1].

[1]Curtin Corrosion Centre, WA School of Mines: Minerals, Energy and Chemical Engineering, Curtin University, Kent Street, Bentley, WA 6102, Australia.

[2]Woodside Energy Ltd., Perth WA 6000. Australia

SUMMARY: Microbiologically influenced corrosion (MIC) can have devastating effects on the durability of oil production equipment and pipelines. Three biocides commonly used in the oil industry were comparatively studied in terms of their effectiveness against biofilm cell viability and biofilm structure and. Bacteria from field-produced water were grown as biofilms on carbon steel coupons in CDC reactors under anaerobic conditions. After two weeks of growth, biofilms were exposed for 4 hours to the different biocides at a concentration of 250 ppm. Changes in cell concentration, microbial activity, and biofilm structure after treatment were used to assess biocidal efficiency. Results indicated that the biocide containing THPS and ADBAC mixture was the most effective biocide for the treatment of mature biofilms. This investigation highlight the importance of the biocide composition for an effective biofilm inhibition that can prevent MIC.

Keywords: Biofilm, Non-oxidising biocides, Oilfield, Steel, MIC

MAINTENANCE OF CONCRETE WHARF STRUCTURES: PORT OF NEWCASTLE CHANNEL & DYKE BERTH DOLPHINS

O. Gasior[1], B. McGuiness[1]
[1]Marine and Civil Maintenance Pty. Ltd

SUMMARY: The Channel Berth, located within the Carrington Precinct of Newcastle Port, is used exclusively for Cruise Ships visiting Newcastle. The structure consists of three dolphins supported by steel piles with interconnecting walkways, with the reinforced concrete deck supported by transverse and longitudinal beams.

To the north of the Channel Berth lies the Dyke Berth 1, which consists of five dolphins, with independent reinforced concrete slabs supported by steel piles.

Corrosion of the reinforcement has led to the concrete's deterioration, particularly evident within the Channel Berth transverse and longitudinal beams. Both the Channel and Dyke Berth dolphins are subject to significant tidal conditions. The undersides of the structures are often inaccessible during high tide events, which presents a considerable challenge to overcome during this project delivery.

Each asset (i.e Dyke Berth or Channel Berth) required a different repair strategy based on the asset's risk profile. The repair treatments included concrete patch repair of the deck slab, soffits and beams, in some instances, the installation of sacrificial anodes in the repairs, and application of silane impregnation to general surfaces. This paper describes as a practical case study the rehabilitation of all the Channel Berth dolphins and two of the Dyke Berth dolphins under active operational conditions during 2018 and 2019.

Keywords: Reinforcing Corrosion, Seawater, Concrete Repair, Maintenance

PROGRESS TOWARDS UNDERSTANDING THE KEY FACTORS GOVERNING PIT STABILITY

Ke Wang, Mobin Salasi, Mariano Iannuzzi
Curtin Corrosion Centre, Western Australia School of Mines (WASM), Faculty of Science and Engineering, Curtin University, Perth, WA, Australia, 6102

SUMMARY: The pit stability product—defined as the product of pit depth and dissolution current density (x·i)—was first introduced by Galvele as the criterion indicating the conditions that sustain a critically acidic solution within the pit. The one-dimensional (1D) electrode is the most commonly used experimental configuration for determining the pit solution chemistry and pit stability product. In our current works, a 'sandwich' like 1D pit electrode was developed, which enabled the in-situ and ex-situ visualization of pit depth. A methodology that avoids lacy cover formation was proposed to pre-dissolve a Type 316L stainless steel 1D electrode to certain depths. Based on the developed methodology and mathematical models, it was found that 1D Fick's law of diffusion could not be used to estimate the pit stability product under a salt film because electro-migration had a measurable contribution (67%) to the dissolution current, which is the first time to be reported in the literature. Although the diffusion coefficient of metal cations decreased with an increasing concentration inside the pit, it could be replaced by a constant diffusion coefficient to estimate the pit stability product, defined as an equivalent diffusion coefficient in our work. Later, a more comprehensive mathematical model, which included hydrolysis reactions and activity, was built to predict the local chemistry within the pit cavity.

Keywords: Stainless Steel, One-dimensional pit, Localised Corrosion, Modelling, Pit Stability Product, Salt film.

On the application of artificial intelligence in the field of corrosion

N. Birbilis, S. Zhao, B. Burton, J. Mangos, Z. Li

College of Engineering and Computer Science, The Australian National University, Acton, A.C.T.,

Presenter email address: nick.birbilis@anu.edu.au

Abstract

Broadly speaking, artificial intelligence (AI) is often defined as the intelligence demonstrated by machines, as opposed to the natural intelligence displayed by humans or animals. The past few years has seen the application of several categories of AI across many disciplines. Specifically in the domain of engineering, and the fields of materials engineering and corrosion engineering, the prospects of AI are abundant. For example, *machine learning* (ML) has been utilised in the exploration of new corrosion resistant materials and in data interpretation. *Computer vision* (CV) has been used to detect the presence or absence of corrosion, without the need for humans in the loop. The application of *natural language processing* (NLP) has also been explored – applied towards answering some of the unique challenges in materials engineering. Herein, examples will be given from practical and 'real world' applications of artificial intelligence in the field of corrosion engineering.

Keywords: corrosion, machine learning, computer vision, natural language processing,

CORROSION AT THE INTERFACE

R.E. Melchers
Centre for Infrastructure Performance and Reliability,
The University of Newcastle, Australia

SUMMARY: Corrosion on the interface between a metal alloy, such as steel, in contact with a wet, permeable non-metallic medium is of considerable practical interest. Examples include exposure to water or the atmosphere or to concrete as for steel reinforcement bars, to soil as for buried cast iron or steel pipes, to deposits of some type as in under-deposit corrosion, to insulation, to protective coatings or macro- or micro-biological agents. In all cases corrosion initiation depends on the characteristics of the interfacial zone both of the metal and of the medium and on spatial variability. For (near-)homogeneous semi-infinite media with good interfacial contact, pitting, crevice and general corrosion of the metal will be controlled largely by the metal (micro-) characteristics, including its inclusions, imperfections and surface roughness. In other cases these may be overshadowed by the macro-characteristics of the medium and the degree of interfacial contact, possibly with severe resulting corrosion. In all cases, however, any longer-term corrosion is dominated by the build-up of rusts at the interface and these govern the long-term corrosion rate. For media of finite thickness, diffusion issues and material deterioration also may be involved. The practical implications are outlined. It is argued that with the presence of a suitable medium it is possible to achieve negligible long-term corrosion but only if certain practical actions are taken.

Keywords: Steel, Soil, Concrete, Deposits, Homogeneity, Long-term.

CLIMATE SMART DURABILITY

S.Furman[1]
[1]AECOM Australia, Melbourne

SUMMARY:

Climate smart durability planning is a process that embeds climate change, life cycle embodied carbon and project sustainability goals into the current durability planning processes. Effective climate smart durability requires collaboration between disciplines and most importantly reliable information. But this information will continually change as climate change data is updated, technologies develop, and research produces new low carbon materials. Therefore, climate smart durability processes will need to be continually updated to optimise material selection to achieve project sustainability targets and actively contribute to company and government targets for net zero carbon emissions. This paper aims to present some of the currently available information that can be utilised in climate smart durability planning.

Keywords: Durability, climate change, embodied carbon, sustainability, net zero carbon.

MICROBIALLY INFLUENCED CORROSION – REFLECTIONS AND FUTURE TRENDS

S.A. Wade
School of Science, Computing and Engineering Technologies,
Swinburne University of Technology, Australia

SUMMARY: Interest in, and the general recognition of, microbially influenced corrosion (MIC) as an important form of structural degradation has rapidly increased over the last few decades. In this presentation I will reflect on key points I raised when I first started working on this topic approximately 16 years ago and discuss what, if anything, has changed since then. This will include several examples of MIC research undertaken by the teams that I have collaborated in over the years. Finally, I will provide some comments on future opportunities in MIC research: there's so much more to do on this exciting and challenging form of corrosion!

Keywords: microbially influenced corrosion, sulfate reducing bacteria, cathodic protection, biocides

AN ELECTROCHEMICAL AND SURFACE STUDY OF 2-AMINO-4-METHYLTHIAZOLE AS CORROSION INHIBITOR FOR GALVANISED STEEL CORROSION IN NACL SOLUTION

Qiushi Deng[1], Steffen Jeschke[1], Billy J. Murdoch[2], Guang Zeng[1], Philipp Eiden[3], Jan N. Gorges[3], Patrick Keil[4], Xiao-Bo Chen[1] and Ivan Cole[1]

1 School of Engineering, RMIT University, Melbourne, VIC 3000, Australia 22

2 RMIT Microscopy & Microanalysis Facility, RMIT University, Melbourne, VIC 3000, Australia

3 BASF SE, Carl-Bosch-Strasse 38, Ludwigshafen am Rhein 67056, Germany

4 BASF Coatings GmbH, Münster 48165, Germany

SUMMARY: Corrosion inhibitory performance of 2-amino-4-methyl-thiazole (2AT-Me) towards galvanised steel (GS) was examined in 0.1 M NaCl solution (pH 7) through electrochemical and surface analysis. Potentiodynamic polarisation scan (PDS) results reveal that 2AT-Me acts as a mixed-type inhibitor passivating both anodic and cathodic corrosion kinetics. Long-term linear polarisation resistance (LPR) test for 7 days indicates that 2AT-Me possesses sustainable inhibition performance with pronounced polarisation resistance (R_p) values as compared to the uninhibited counterpart. Furthermore, surface analysis using X-ray photoelectron spectroscopy (XPS) and scanning electron microscopy (SEM) techniques was carried out to determine the adsorption state and film formation of 2AT-Me molecules. The results suggest that sulphur heteroatom and attached amino functional group act as adsorption centre facilitating the subsequent inhibitor film formation to inhibit corrosion of GS surface.

Keywords: Galvanised steel, Corrosion inhibition, Electrochemical characterisation, Surface analysis

LABORATORY TESTING OF ZINC-RICH PAINTS: WHAT CAN YOU LEARN?

R A Francis
R A Francis Consulting Services, Ashburton, Vic, Australia

SUMMARY: Investigations into the performance and protection mechanisms of protective coatings will typically require accelerated exposure or electrochemical tests or both to be carried out. A number of such tests have been developed over the years which range from quick and simple exposure tests to complex electrochemical investigations. This paper reviews the most popular tests used in investigations, specifically into the performance zinc-rich paints, mainly directed at coating users and specifiers rather than researchers. A brief description of the methods, how the results may be presented, examples and advantages and limitations are given.

Keywords: Zinc-rich paints, accelerated testing, laboratory testing.

ZINC-RICH PAINTS: EXPLORING PROTECTION MECHANISMS

R A Francis
R A Francis Consulting Services, Ashburton, Vic, Australia

SUMMARY: Zinc-rich paints are the protective coating of choice as primers and single coat systems for many aggressive environments. They provide sacrificial protection to the steel substrate and it is this property that sets them apart from other coating types. There must be some sort of alternative protection to explain their performance for long-term protection when cathodic protection may be lost, but there have been many differing opinions and theories provided over the years. This paper reviews some of the field exposure, accelerated testing and electrochemical investigations that have been carried out on zinc-rich paints over the years to attempt to provide a clearer understanding of how they provide protection.

Keywords: Zinc-rich paints, sacrificial protection, zinc content, inhibition.

CORROSION OF TRANSFORMER LID RESULTS IN EXTENSIVE FIRE – A CASE STUDY

R. Jeffrey[1], G Sussex[2]

[1]Pacific Testing,

[2]Sussex Materials Solutions

SUMMARY: Discussed herein are the root causes and consequences of the corrosion of a high voltage transformer lid in south-east Australia. During an extended heatwave some years ago, where temperatures soared well into the 40's, an electrical overload caused the fuse of transformer to "blow" The hot fuse was ejected from the transformer and landed in surrounding dry grass initiating a fire which subsequently spread and caused significant damage to nearby storage sheds and their contents. Investigations revealed significant corrosion to the lid of the transformer including several perforations which had allowed the ingress of water, reducing the dielectric properties of the transformer oil. The loss of insulating properties of the oil allowed the coils to "short" during an overload resulting in the blown fuse and consequent fires. The geometry of the transformer, the corrosivity of the environment, the difficulty of undertaking a thorough inspection and the age of the equipment all contributed to the corrosion of the utility and subsequent damage. These factors are discussed.

Keywords: Corrosion, high voltage transformer, ponding

RE-EVALUATING THE MECHANISMS DRIVING LONG-TERM CORROSION OF THE 'TITANIC'

Robert E. Melchers[1]
[1]Centre for Infrastructure Performance and Reliability, The University of Newcastle, Australia

SUMMARY: Most current opinion supports the idea that microbial activity drives the deep water immersion corrosion of the ill-fated passenger ship Titanic, sunk in 1912. It is located in very low oxygen, low temperature seawaters. Herein it is proposed that the observations of rusticles are consistent with physio-chemical corrosion that may or may not involve microbiologically influenced corrosion. It also is consistent with the corrosion rate observed over a 6-year experimental trial at the site and other site-specific information. Already available models indicate that long-term corrosion occurs predominantly under anaerobic conditions at the corrosion interface. The corrosion products so formed can, even under very low dissolved oxygen conditions, oxidize slowly on their outer surfaces to maghemite. This process also releases ferrous ions into the surrounding environment where they are likely to form the rust products observed by divers and marine archaeologists. Conversely, there is no evidence to date of the sufficient presence, at the level of the wreck, of dissolved inorganic nitrogen, the primary macro-nutrient necessary for the occurrence of microbiological influenced corrosion in seawater.

Keywords: Corrosion, ship wrecks, rusts, oxidation, microbiological activity.

DEGREE OF BURIAL ON THE CORROSION OF MILD STEEL IN SAND AND SEAWATER

R.B. Petersen, R.E. Melchers, M.M. Hossain, I.A. Chaves
Centre for Infrastructure Performance and Reliability, The University of Newcastle, Australia 2308

SUMMARY: There are many practical scenarios where mild steel is corroded in the presence of both seawater and inert particles, such as sand or coal. Corrosion under these conditions is a form of under-deposit corrosion, with the rate and extent of corrosion a function of the degree of burial. This paper presents experimental work aimed at studying this behaviour. Mild steel coupons were buried in sands (either completely, partially or not buried) and immersed in seawater for up to 1 year. Completely buried coupons displayed the least corrosion, presumably due to a decrease in oxygen availability to the metal surface. Settling of marine sediments next to the surface of the partially buried coupons provided conditions for crevice corrosion. The progression of corrosion pit depths and mass loss with time are consistent with the initial stages of the bi-modal trend for steel corrosion.

Keywords: Mild steel, Under-deposit corrosion, Mass loss, Pit depths, Experimental study

PARTICLE SIZE ON THE CORROSION OF MILD STEEL IN PARTICULATE MEDIA AND SEAWATER

R.B. Petersen, R.E. Melchers, M.M. Hossain, I.A. Chaves
Centre for Infrastructure Performance and Reliability, The University of Newcastle, Australia 2308

SUMMARY: Steel structures are known to corrode in the presence of seawater and inert particles (sands, rusts, coal, iron ore, etc) of different sizes. Previous work by others has shown that particle size can influence corrosion mass loss rates. However, it remains to be seen whether particle size has an influence on the long-term corrosion mass loss and pit depths under fully immersed seawater conditions. This paper presents experimental work aimed at studying this behaviour. Mild steel coupons were buried in three different particulate media (sand, a coarse silicon carbide, and a fine silicon carbide) and immersed in seawater for up to a year. Pitting depth and extent were found to be independent of particle size. Mass loss was found to increase with an increase in particle size. The progression of corrosion pit depths and mass loss with time are consistent with the initial stages of the bi-modal trend for steel corrosion.

Keywords: Mild steel, Under-deposit corrosion, Particle size, Mass loss, Pit depths, Experimental study

WHAT IS SHARP ANGULAR PROFILE

R. Dankiw[1], D. Fosdike[2]
Principal Inspection Consultant of Asset Inspection Consultants Pty Ltd[1], Adelaide, Australia, Materials and Welding Engineer of BAE Systems Maritime[2], Adelaide, Australia.

SUMMARY: This paper attempts to define in technical and non-technical terms how we define and assess a sharp angular profile. Sharp angular profile is commonly required by many paint manufacturer's technical data publications relating to surface preparation requirements, especially in severe environments. Defining sharp angular profile in measurement/mathematical terms is difficult but is relatively simple when using visual standards such as surface comparators. Surface comparators replicate the surface profile produced under controlled conditions depending on the type and size of abrasive media used. There are also new measurement techniques that replicate the surface using portable optical recording devices. This can be useful in recording the achieved anchor pattern, particularly if there is blend/mixture of abrasives used. There is some subjectivity when it comes to visually assessing sharp angular profile that is produced by a mixture of different types of abrasives used.

Keywords: Surface profile, anchor pattern, abrasive blasting, abrasives, adhesion.

NEW HIGH PRODUCTIVITY SILENCED NOZZLE

S. Simmons

BlastOne International

SUMMARY:

Research, conducted in conjunction with the University of Queensland, focused on Abrasive Blast Nozzle performance and related geometrical characteristics has resulted in the development of a new silenced blast nozzle that reduces noise generated by up to 16dBA and nozzle thrust by up to 45%.

Keywords: Abrasive Blasting, Blasting Noise Reduction, Blast Nozzle Thrust Reduction, GMA Garnet, Abrasive Blasting Safety

SUPPORTING INDUSTRY ONE COAT AT A TIME
THE EVOLUTION OF APAS-PCCP

T Gramlick, Manager Verification Services – Infratech, CSIRO
E Stepanova, Executive Officer PCCP, CSIRO

SUMMARY:

Over many decades, governments, specifiers, procurers and applicators have relied on the Australian Paint Accreditation Scheme (APAS) and its partner, the Painting Contractor Certification Program (PCCP) to support conformity of products and services in their particular sector. These schemes have been successful and the stakeholders who have supported them can be proud that this industry led approach has delivered a trusted market which is not tarnished with the perception of 'dodgy' products currently evident in other parts of the building sector.

We recognise however, that such schemes cannot remain static and to meet industry's future needs and ensure conformity of products and services they, along with the CSIRO, need to evolve and change. The CSIRO works with key stakeholders to better understand their concerns with the schemes and explore future options and opportunities. Ongoing consultation with organisations has firmly reinforced the need for ongoing change and change is well underway.

This paper outlines the scope of products and work included in the programs and highlights the inextricable link between product conformity and its correct selection and application. It also identifies the issues and the opportunities with the APAS-PCCP schemes and provides a progress report on the work underway with a glimpse into new and aligned initiatives for the near future.

Keywords: Paint, protective coating, APAS, PCCP.

REHABILITATION OF CULVERTS THROUGH SPIN CASTING STRUCTURAL LINERS.

W J Thomson[1]

[1]PFP Systems Pty Ltd

email address: wayne@pfpsystems.com

SUMMARY: Culverts are the often-unseen bridges of transport infrastructure. They are traditionally reinforced concrete, galvanized steel and sometimes aluminium formed and assembled pipes. They generally provide a bridging structure for occasional water beneath roads, railways, paths and tracks. Over time, these structures deteriorate due to ground movement, induced damage, erosion and corrosion. Traditionally, culvert rehabilitation requires jacked replacement pipes, in-situ concrete forming and trench and replace methods. However, a further method, that of centrifugally casting a structural liner composed of a specially designed repair mortar is often a viable and economical solution. Depending on the design, this method can extend the life of culvert for up to 100 years.
This paper details spin casting methodology, presents several cases of deteriorated culverts and examines the process by which they were restored to their structural loading capacity.

Keywords: Culvert rehabilitation, bridge repair, corrosion prevention; road, storm water, sewer infrastructure.

CORROSION TESTING AND RISK ANALYSIS FOR DESIGN OF REPAIR OPTIONS FOR REINFORCED CONCRETE, INCLUDING CATHODIC PROTECTION SYSTEMS

Z. Wang[1], J. Dyson[1]
[1]BCRC, Sydney

SUMMARY: Over time, more and more concrete structures approach the end of their safe service life. The process of accurately assessing a concrete structure's condition and remaining service life, to enable efficient repair and protection measures to be designed to extend a concrete structure's useful life, is of critical importance to owners. This paper discusses valuable field and laboratory concrete testing methods, including non-destructive testing and destructive testing methods to analyse the likelihood of chloride-induced and carbonation-induced corrosion in reinforced concrete structures.

Traditional and innovative concrete repair options include patching, crack injections, concrete sealers, protective coating, carbon fibre strengthening, and various types of Impressed Current Cathodic Protection (ICCP) and Sacrificial Anode Cathodic Protection (SACP) systems will be discussed. This will include the differences, and possible advantages and disadvantages of various CP systems in different applications. Case studies will be presented for remedial design and specification of swimming pool, concrete slab with magnesite flooring and car park.

Keywords: Concrete, Testing, Corrosion, Repair, Cathodic Protection

AUTHOR INDEX